My First Science Books
A CRABTREE SEEDLINGS BOOK

MY SCIENCE TOOLS

Julie K. Lundgren

A Crabtree Seedlings Book

Crabtree Publishing
crabtreebooks.com

We use science **tools** to explore, ask, and learn.

How long is it?

0 INCHES 1 2 3 4 5

Use a ruler to find the **length**.

How heavy is it?

Use a **scale** to weigh it.

I want to look closer.

Use a **hand lens** to zoom in.

Use a **microscope** to zoom in even more.

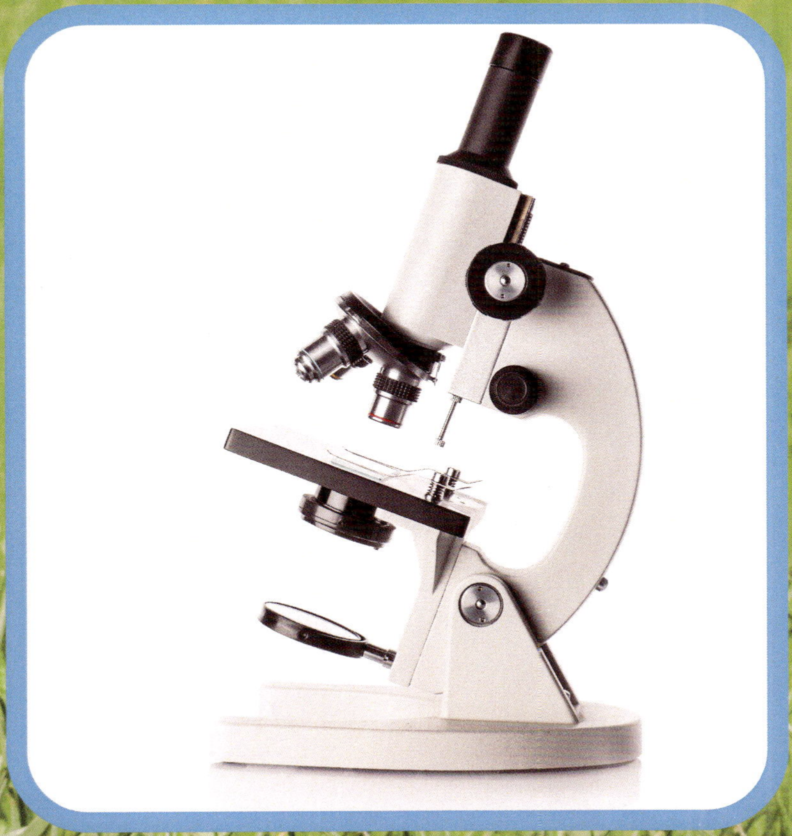

Use a **tablet** to store what you learn.

Science tools help us see, know, and think about the world around us.

Glossary

hand lens (HAND LENZ): A hand lens is a tool with a curved glass that we look through to make things look bigger.

length (LENGKTH): Length is how long something is from one end to the other.

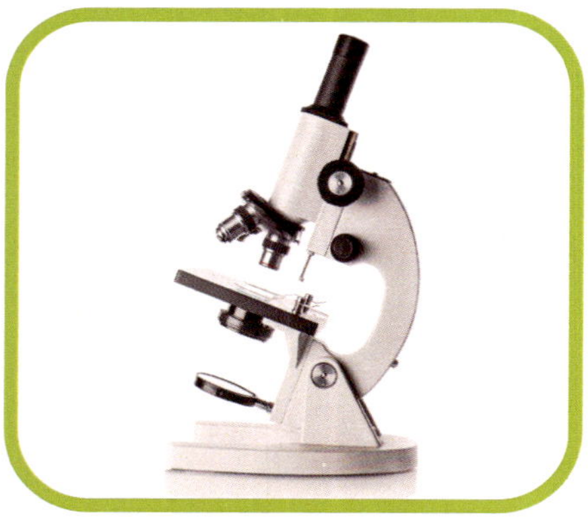

microscope (MIKE-ruh-skope): A microscope is a tool with a strong lens that lets us see tiny things up close.

scale (SKALE): A scale is a tool we use to weigh things and find out how heavy they are.

tablet (TAB-let): A tablet is a small computer that is easy to take with us, and is used to store and share our notes and ideas.

tools (TOOLZ): Tools are things we use to help us do jobs.

Index

School-to-Home Support for Caregivers and Teachers

Crabtree Seedlings books help children grow by letting them practice reading. Here are a few guiding questions to help the reader with building his or her comprehension skills. Possible answers are included.

Before Reading

- **What do I think this book is about?** I think this book is about the tools people use to help them learn science.
- **What do I want to learn about this topic?** I want to learn the names of different science tools.

During Reading

- **I wonder why...** I wonder how heavy the rabbit is.
- **What have I learned so far?** I have learned that rulers, scales, hand lenses, microscopes, and tablets are science tools.

After Reading

- **What details did I learn about this topic?** I learned that hand lenses and microscopes help people see something closer.
- **Read the book again and look for the vocabulary words.** I see the word ***length*** on page 7 and the word ***scale*** on page 10. The other vocabulary words are found on pages 22 and 23.

Crabtree Publishing

crabtreebooks.com 800-387-7650

In Canada: We acknowledge the financial support of the Government of Canada through the Canada Book Fund for our publishing activities.

Hardcover 978-1-4271-3029-7
Paperback 978-1-4271-3040-2

Printed in Canada/062024/CP20240617

Published in Canada
Crabtree Publishing
616 Welland Avenue
St. Catharines, Ontario
L2M 5V6

Published in the United States
Crabtree Publishing
347 Fifth Avenue
Suite 1402-145
New York, NY 10016

Written by: Julie K. Lundgren

Photo Credits: Cover © A3pfamily; page 2-3 © Rawpixel.com; page 4-5 © Dave Pot; page 6-7 ruler © Kitch Bain, Chawalit Chanpaiboon, gecko © Chawalit Chanpaiboon; page 8 © ravel, page 9 © Pair Srinrat, page 10-11 © Tinxi; page 12-13 © A3pfamily, page 14-15 © Pavel.Proc; page 16-17 © Bernad; page 18-19 © Yuliia V; page 20-21 © Rawpixel.com All photos from Shutterstock.com

Library and Archives Canada Cataloguing in Publication
Available at the Library and Archives Canada

Library of Congress Cataloging-in-Publication Data
Available at the Library of Congress